Genetic Engineering

by

D. Gareth Jones

Associate Professor of Anatomy and Human Biology,
The University of Western Australia, Perth

GROVE BOOKS

BRAMCOTE NOTTS.

CONTENTS

		Page
	Preface	3
1.	Genetics and Society	4
2.	Genetic Control	8
3.	Techniques of Genetic Control	12
4.	Genetics and the Human Predicament	19

First Impression September 1978

ISSN 0305 4241

ISBN 0 905422 42 2

PREFACE

With the success of man's past efforts at controlling the myriad diseases that have scourged him from his external environment, he has arrived at a point where he can reflect on deficiencies in his internal environment—his own biological make-up. This opportunity to meditate upon the quality of his genetic inheritance has opened up ways in which it can be modified.

Already man has delved deeply into his genetic constitution, so that he can now influence the course of certain genetic diseases. This, however, is only the beginning. Far more fundamental and radical steps are in the offing, and it is these that are raising far-reaching questions about man's nature, and the extent to which his expectations of human life can and should be altered. Alongside these basic issues about human existence go more immediate ones, concerning the nature of marriage and family life, the degree of freedom individuals should be allowed in the choice of a mate or in the production of offspring, and the type of individuals who should be permitted into the human community.

Genetic advance has these and many other theological, ethical and social issues implicit within it. Basic to them all, however, is the even more fundamental one of quality control. Medicine throughout the centuries has striven to improve the immediate well-being of individuals. This has had repercussions for the quality of life of whole societies, and even of whole countries. What the genetic revolution is bringing in its wake is the conscious desire for improvement in quality. But why should genetic control be sought, and what are its goals? How may it affect our view of man and how may it influence our appreciation of the value of individual life? And, an ever-present fear today, in what respects may it be misused?

The possible dangers of genetic manipulation appear to have convinced a wide section of the community that biologists are rapidly moving out of an age of innocence. This is because they give the impression of being on the threshold of understanding the genetic programme of living organisms, as a prelude perhaps of the creation of new programmes and hence new varieties of organisms. Arthur Rosenfeld, a popular writer on genetics, waxes eloquent that there lies before us 'the control of life. All of life, including human life. With man himself at the controls. Also coming: a new Genesis—The Second Genesis. The Creator, this time around—man. The creation—again, man. But a new man. In a new image.'[1]

Before dismissing such poetic imaginings as journalistic nonsense, contemporary genetics is indeed ushering in a biological revolution which has the potential of allowing man to exercise control over his very existence and destiny as a biological being. It may affect what man is and what he is becoming; it may enable him to alter himself and his expectations of human life in radical and highly novel ways.

Whatever our reactions to the possibilities of genetic manipulation, both imminent and distant, both realistic and highly-imaginative, we are confronted by one stark truth—glib answers are not likely to be helpful. We cannot afford to tackle the issues posed by genetic advance solely on the basis of past authoritarian answers. At all levels of this debate, it is essential that a Christian contribution be developed and advocated, but this will require that a bridge be built between contemporary genetic issues and relevant ethical principles. My aim in this booklet is to attempt to show how such a bridge might be constructed.

[1] Rosenfeld, A., *The Second Genesis* (Arena Books, New York, 1972) p.21.

1. GENETICS AND SOCIETY

Although the brunt of popular debate about genetics centres on genetic engineering, this is but the futuristic tip of a much more general area of concern, namely, the magnitude of genetic disease in today's world.

The incidence of genetic disease is illustrated by the fact that three per cent of all children have some genetic defect at birth, while a much higher incidence of conceptions miscarry due to identifiable chromosomal abnormalities. Indeed, of the 25 per cent of all conceptions which fail to survive to birth, a third are associated with chromosomal errors. Another illustration of the importance of genetic disease is the fact that about 1,600 human diseases have a genetic origin. Of these, conditions such as cystic fibrosis and sickle-cell anaemia are quite common, while approximately 50 per cent of all cases of both congenital blindness and congenital deafness have a genetic base.[1] When the probability that the incidence of genetic disease in the human population is increasing is considered, the amount of present and future suffering emanating from this source becomes enormous.

The major causes of genetic defects are deleterious genes and chromosomal abnormalities.[2] About ten per 1,000 live births in European populations carry harmful genes responsible for genetic disease, and five per 1,000 suffer from chromosomal abnormalities. Of the large number of genetic defects, some are relatively common and can be detected by genetic screening. The principal such defects are Down's syndrome (Mongolism), cystic fibrosis, phenylketonuria (PKU), sickle-cell anaemia and Tay-Sachs disease. Of the metabolic diseases caused by deleterious genes, glucose-6-phosphate dehydrogenase deficiency is a notable example.

Perhaps the best-known of these defects is *Down's syndrome,* with an incidence of one in 600-900 live births. This is due to an excessive number of chromosomes, most cases having three number 21 chromosomes in each cell, instead of the normal two.[3] The social importance of this condition stems from its relatively high incidence and the fact that it is far more common in older mothers than in younger. Furthermore, many of those afflicted are severely mentally-retarded, while the remainder are less severely although definitely retarded. Seven thousand children suffering from Down's syndrome are born each year in the United States, these children comprising as many as 30 per cent of those school-aged children having severe mental handicap. As Down's syndrome can be detected *in utero,* an affected foetus can, if desired, be aborted.

1 For this information I am indebted to Professor Charles Birch's article 'Genetics and moral responsibility' in Birch, C. and Abrecht, P. (ed.), *Genetics and the Quality of Life* (Pergamon Press, Sydney, 1975) pp.6-19.

2 Genes are the basic units of heredity and are carried on chromosomes. Each gene consists of deoxyribonucleic acid (DNA). Forty-six chromosomes occur in each human cell, each one serving as a carrier of genetic material.

3 For a fuller discussion, see Jones, A. and Bodmer, W. F., *Our Future Inheritance: Choice or Chance?* (Oxford University Press, London, 1974) Chapter 4.

This detection is accomplished by means of *amniocentesis.* In this a sample of the amniotic fluid is removed from around the foetus at about the sixteenth week of pregnancy. Because the foetus sheds cells into this fluid, analysis of the fluid enables the diagnosis of defects caused by some defective genes and also by abnormalities in the structure or number of the chromosomes. At present only a few of the immense variety of genetic diseases can be detected in this way, although the number is increasing steadily. A positive diagnosis of Down's syndrome or some other genetic/chromosomal defect by amniocentesis opens the way for a therapeutic abortion. The inevitable ethical considerations inherent in this procedure will be analyzed in a later chapter.

Other conditions involve defects, not of chromosomes, but of particular genes making up the chromosomes. For instance, *haemophilia* and *Duchenne muscular dystrophy* are caused by abnormal genes on one of the sex chromosomes, and hence are *sex-linked diseases.* In general only males suffer from them, although they are transmitted through females. Important as these two conditions are, they are relatively uncommon. For instance, in Britain where there are of the order of 800,000 births a year, approximately 160 cases of Duchenne muscular dystrophy occur and only 80 cases of haemophilia. One reason why these conditions are significant is that, because of the manner in which they are inherited, their incidence can be lowered by careful genetic counselling.

Other genetic diseases are linked, not with the sex chromosomes, but with one of the 44 other chromosomes in the nucleus of a normal human cell. These fall into two general classes, *dominant* and *recessive* conditions.

A well-known disease caused by a dominant gene is *Huntington's chorea,* which commences in the third or fourth decade of life and leads eventually to severe mental deterioration and finally death. Whenever the offending gene is present, the disease emerges. Each afflicted person has a 50 per cent chance of passing the condition on to his or her offspring, with men and women being equally affected. The high probability of passing on the condition to one's children and the inevitability of early death when afflicted constitute dimensions of tragic proportions.

Far more common than this type of genetic disease, however, are the recessive conditions of which some 800 are currently recognized. Fortunately, most of these are rare. In these conditions the gene responsible for the characteristic is much more common than the disease itself, as two such genes must be present before the disease manifests itself, one coming from the mother and one from the father. Even then the probability of any child of such a union having the disease is only 25 per cent.

In Britain, the most common recessive diseases are *mental deficiency, deafness* and *blindness,* each of which is a consequence of a number of different single-gene defects and, therefore, very difficult to unravel in genetic terms. For instance, one child in every 1,300 births in Britain is mentally defective. Far more is known about those recessive diseases due to one particular gene of which the most common example is scytic fibrosis of the pancreas, with an incidence of one affected child in every 2,000 in most Western countries.

In view of the social importance of the recessive genetic diseases, I shall elaborate on three of them which have been investigated in considerable depth. The first of these is *sickle-cell anaemia,* in which individuals carry two abnormal genes and suffer from an abnormality of haemoglobin, dying in most cases before the age of 20. In parts of West Africa, one child in 64 is affected, with one in 500 black children similarly affected in the United States. Besides these affected individuals, about two million people carry just one abnormal gene. These are the 'carriers', who are clinically normal and who, in parts of Africa, manifest an increased resistance to malaria. Carriers can be identified by screening programmes, although this does not apply to the foetus.

The ethical and practical problems concerning what is to be done with the knowledge acquired from screening constitute a major obstacle in treating this condition in a radical way.

Sickle-cell anaemia could be eradicated by preventing carriers from mating with one another. If this were implemented, it would still leave the carriers 75 per cent of the population from which to choose their mates.[1] The principal issue emanating from this possibility is whether such a restriction of freedom could or should be applied to only one section of the population. As it is the black population alone which is affected, this would be a particularly contentious proposal in a country like the United States.

Phenylketonuria (PKU) is another recessive condition which, in European populations, has a frequency between one in 10,000 and one in 20,000. PKU represents an extremely interesting condition from a medical standpoint because, although it cannot be cured, its clinical manifestations can be controlled. The genetic deficiency in this instance leads to the body's inability to produce an essential enzyme, with the result that toxic products accumulate in the body, leading to mental retardation. These consequences can, however, be prevented provided the disease is detected very early in life and the individual is subsequently kept on an appropriate diet. Screening for PKU is now carried out in the majority of births in Britain and the United States.

Conflict in this case amounts more or less to a financial one, in that the costs of a wide-scale screening programme may outweigh the costs of keeping affected individuals (assuming no screening and no early dietary treatment) in mental institutions for the rest of their lives. On the other hand, the cost of screening and treatment per affected individual is only about 16 per cent that of the lifelong maintenance of an untreated individual in an institution. It may seem callous even to raise the question of money when a method of successful treatment is known. This may be so, but the

1 See Bodmer, 'W. F. Biomedical advances: a mixed blessing?' in Harre, R. *Problems of Scientific Revolution* (Clarendon Press, Oxford, 1975) pp.25-41.

cost-benefit ratio advantage for a screening programme decreases as the frequency of the disease decreases. A point must be reached, therefore, where the potential benefits of the screening can no longer justify the financial outlay. Even when screening is used, as for PKU, some apparently positive results are not, in fact, due to PKU. The possibility that then has to be considered is whether these 'false-positive' individuals will suffer from being 'treated' with the special diet.

Tay-Sachs disease, like PKU, is an inborn error of metabolism—being caused by a recessive gene and having a one-in-four chance of appearing in children of carriers. It has a frequency of one in 2,000 among Askenazi Jews, but a very much lower frequency in other populations. Infants suffering from it develop progressive mental deterioration, followed by blindness and paralysis, and death by the age of about three years. Because both carriers and affected foetuses can be identified, normal children can be guaranteed to matings between carriers by aborting affected foetuses. Screening programmes involve wives, husbands, and finally foetuses, and there can be no doubt that, in cost-effective terms, they are highly successful among the Askenazi Jews in the United States. The one proviso is that *in utero* detection of the disease must be followed by therapeutic abortion if the programmes are to be of any avail. An unwillingness to go through with subsequent abortion negates the rationale of the schemes, raising yet again the crucial significance of therapeutic abortion for their success and highlighting its pivotal position as an ethical question.

2. GENETIC CONTROL

The approach to genetic problems considered in the last chapter is the province of *prenatal control* or *negative eugenics*—negative because it involves the elimination of defective genes (and hence the prospective possessors of these genes) from the population. While negative eugenics is not a radical approach to the genetic constitution of man, and while it does not actually change individuals, its impact on society is an important one and its potential for changing the genetic make-up of a community is far from negligible.

Ethical issues intrude far more than often suspected into negative eugenics, because man is forced even at this level into a situation of responsibility. C. S. Lewis placed this responsibility in perspective with his assertion that 'each new power won *by* man is a power *over* man as well.'[1] With slightly different emphasis but much the same thrust, a Norwegian Marxist theologian has expressed our dilemma thus: 'Freedom from the tyranny of nature generally means coming under the tyranny of men. The more technology makes us free from the pressures of nature, the heavier must the social and economic pressure of some people fall on the rest of us.'

The challenge of negative eugenics has been provocatively expressed by Professor Bentley Glass.[2] According to him, 'the right of individuals to procreate must give place to a new paramount right: the right of every child to enter life with an adequate physical and mental endowment.' The consequence of such a position is that the diagnosis of genetic disease in a foetus carries with it the moral responsibility to minimize the birth of defective children. And so emerges Glass's now-famous remark that advances in human genetics should usher in the day when the paramount right of every child is to be born with a normal, adequate, hereditary endowment. Such a stance assumes that biological quality is the paramount goal of genetics, and hence is to be the overriding directive of human endeavour. The ethical basis for this will be considered in another chapter.

The complexity of the decisions that have to be taken by parents before consenting to foetal diagnosis and consequent abortion has been outlined by the Church and Society sub-unit of the World Council of Churches.[3] This assessment, while a helpful guide to the issue, accepts the legitimacy of therapeutic abortion and, therefore, mirrors an approach incorporating abortion as an indigenous part of the procedures. According to this outline, any detriment resulting from the birth of the foetus outweighs the potential benefits. The criteria suggested include: (i) the severity of the genetic disorder and its effects on the possibility of a meaningful life; (ii) the physical, emotional and economic impact on family and society; (iii) the availability of adequate medical management and of special educational facilities; (iv) the reliability of diagnosis; (v) the recognition that an individual genetically defective in one respect may be superior in others; (vi) the increase in the load of detrimental genes in the population that may result from the reproduction of carriers of genetic diseases.

If these criteria are accepted as useful guides to negative eugenics, they demonstrate that actual or probable genetic disease of a foetus cannot

1 Lewis, C. S. *The Abolition of Man* (MacMillan, New York, 1965) p.70.

2 Glass, B. 'Science: endless horizons or golden age?' in *Science, 171,* 23-29 (1971).

3 Report of Church and Society sub-unit of the World Council of Churches: 'Genetics and the quality of life' in *Study Encounter,10,* 1-16 (1974).

be approached in a simple black-and-white manner. The biological health of a prospective child is not the only guide to the course of action to be taken, as Glass would have us believe. A number of questions are relevant, if human considerations are not be to overlooked. For instance: is a society justified in denying existence to individuals likely to develop a genetic disease? From this it follows that, if decisions are taken to prevent such individuals coming into the world, in whose interests are these decisions? Similarly, it needs to be demonstrated that, if individuals should not be born on the basis of the cost of their future upkeep or on the grounds of the genetic fitness of the human race, whose interests are uppermost—the alleged good of society or that of the progeny?

Even genetic screening raises ethical problems. These focus attention on issues such as the stigmatization of affected individuals, confidentiality and breaches of individual rights to privacy, and freedom of choice in childbearing. Principles for the design and operation of screening programmes should take these issues into account. although they are not readily overcome. While the former two may not involve great ethical principles, they are of utmost relevance to the integrity of personhood.

Positive eugenics or *preconceptive control* has far more radical vistas than negative eugenics, and yet any approach to it depends heavily on present experience with negative eugenics. Control in this instance precedes conception, the intention being to produce an individual who differs in some important respects from the individual who would have resulted without this intervention. In other words, a deliberate effort is being made to produce an individual with 'new' specifications. Hence this is the realm of genetic engineering proper, illustrating as it does positive—as opposed to negative—eugenics.

Genetic engineering has been defined by the American Medical Association in these terms: 'It might be considered as covering anything having to do with the manipulation of the gametes or the foetus, for whatever purpose, from conception other than by sexual union, to treatment of disease *in utero,* to the ultimate manufacture of a human being to exact specifications.'[1]

Preconceptive control invites a great deal of idealistic support, with visions of rectifying genetic and chromosomal aberrations prior to conception. Utilizing these techniques it is envisaged that the amount of human misery will be reduced, and that the struggles of postnatal or traditional medicine will be diminished. According to Joseph Fletcher, 'the ultimate goal of genetic engineering is not to ameliorate the ills of patients prenatally or postnatally, but to start people off healthy and free of disease . . . It aims to control people's initial genetic design and constitution by gene surgery and by genetic design.'[2] This is the essence of *quality control,* the goal of which is to be able to choose who shall live as opposed to who shall not, those selected to live possessing characteristics of which the controllers approve. For Fletcher, whose situation ethics reach unprecedented levels of optimism in the genetic realm, quality control in birth technology should select for intelligence, on the ground that control is human and rational and, therefore, to be espoused.

1 See *J. Amer. Med. Assoc., 220,* 1356 (1972).

2 Fletcher, J., *The Ethics of Genetic Control* (Anchor Books, Garden City, New York, 1974) p.56.

Discussions of this nature are all very well, but tend to ignore the current status of research in genetic engineering. In organisms, such as tadpoles and newts, which have a very simple chromosomal apparatus, one gene has been substituted for another by replacing the original DNA with foreign DNA. In similar organisms whole genes have been transferred from one cell to another suggesting that gene transplants may be a feasible proposition. Another accomplishment is the 'switching on' of particular genes, normally inactive in cells, in order to produce enzymes not normally produced by those cells. For instance, DNA extracted from virulent pneumococci is capable of transforming non-virulent pneumococcus organisms into permanently virulent ones. This involves the replacement of the defective gene by an effective one contained in the extracted DNA, and is the process of *bacterial transformation.* Alongside this can be placed the phenomenon of *transduction,* in which a harmless virus is used as the carrier of a gene from a donor to a recipient possessing only the defective gene. In this way a second gene can be used to replace a defective one.

More recently, the revolutionary aspects of genetic engineering have become yet more apparent with the emergence of a technique for manipulating genes from living organisms. This is the sphere of *recombinant DNA technology,* its benefits and hazards lying in the power it affords to design and create combinations of genes in ways quite different from the slow re-shuffling found in nature. This, in turn, bestows upon man the ability not only to speed up evolutionary change but drastically to alter its direction. As one writer has put it: 'Now at last man has a handle on the force that shaped him.'

This dream (or nightmare) has been made possible by the recent discovery of a class of enzymes known as *restriction enzymes,* which are used by bacteria to recognize and subsequently destroy foreign DNA. As such, they serve as a protective mechanism. However, now that they have been discovered by man, they can be employed to rearrange genetic material in combinations unlikely to occur under natural circumstances. This, in essence, is genetic manipulation, which opens the way to correcting genetic deficiency diseases, providing crop plants with the genes for nitrogen fixation so that they no longer require nitrogen fertilizers, and constructing new strains of bacteria for a range of tasks including the digestion of oil spills.

Of these three possible applications, all would serve man in one way or another, but the one of direct consequence for the nature of man is the first—the elimination of genetic deficiency diseases, providing cures for, among others, myopics, diabetics, haemophiliacs and sufferers from PKU. Such applications of genetic manipulation, however, lie in the future. Of more immediate relevance are the potential hazards of the technique, hazards which a few years ago were considered of such importance as to warrant an embargo on this type of genetic experimentation.

Most of these experiments involve the transference of genetic material to the bacterium *Escherichia coli* which, although a useful experimental organism, also lives in the human throat and intestinal tract. If, therefore transferred genes were to convert this relatively tranquil bacterium into a killer, and were this killer bacterium to escape into the population, an epidemic of unexperienced magnitude could ensue. Another danger is that the new organisms may have acquired the wrong properties, with potentially serious consequences were the new bacterium to escape from the laboratory.

These dangers are potential rather than actual, and guidelines to forestall such dangers were issued by a National Institutes of Health Committee in the United States at the end of 1975. Similar guidelines have since been issued by government bodies in other countries. These are based on physical and biological containment principles, the stringency of the containment levels depending on the likely dangers of particular experiments.

From the Christian perspective, the priorities of this type of research need to be assessed, in terms of its potential benefit to mankind. Science is a God-given way of investigating the natural world, exerting control over it and living out this exercise of authority in a responsible manner. Alongside this must be placed man's sinfulness and the fundamental dilemmas this brings in its wake. Recombinant DNA research probably is dangerous territory, but cannot be regarded as forbidden territory unless we abrogate our God-given responsibility. A system of checks and balances is essential, in order to balance the excesses of man's fallen nature. Not surprisingly, there is no easy solution either in going forward or turning back. To expect the opposite is to fail to understand both man's nature and God's world.

Positive eugenics has a very much longer history than this type of genetic engineering. Nevertheless, the two have much in common regarding their intention. Francis Galton invented the term, 'eugenics', in 1883, the essence of which is to encourage the reproduction of the select. The difficulty has always been to identify the select and, having done this, to promote only those genes considered favourable. Unfortunately, everyone in the population carries and may transmit several lethal genes and as many as a dozen detrimental ones. From this it follows that all known genetic diseases cannot be eradicated without first eradicating all human beings.

Another practical difficulty is to decide for which characteristics selection should be made. Even were everyone to opt for such qualities as loving-kindness, compassion, generosity and courage, the issue would then resolve into the genetic basis of such qualities. Professor R. J. Berry argues that it would be impractical to select for two or more such attributes 'since any responsible genes will be distributed throughout the genome, and could only be concentrated at the cost of imbalancing other developmental systems.'[1] Although it may be possible to improve some characteristics, the inevitable accompaniment may well be mental deficiency and other equally regressive traits.

Positive eugenics, therefore, still has many inherent limitations. What is more, discussions of positive eugenics are elitist, in that they reflect the prejudices of racial, social, political and economic élites. They also assume that biological characteristics are of supreme importance, ignoring the perspective of the individual as a whole person.

It is not difficult to see why positive eugenics, as traditionally conceived, has failed to open up the revolutionary vistas once imagined. Nevertheless, the hope enshrined in it lives on in modified form, both in genetic engineering and negative eugenics, highlighting its humanistic and reductionistic character.

1 Berry, R. J. 'Genetical engineering' in *Christian Graduate 26,* 3-7 (1973).

3. TECHNIQUES OF GENETIC CONTROL

Discussions of genetic advance are of theoretical interest alone until viewed against the background of techniques currently being developed to make it a reality. These techniques in themselves raise fundamental ethical questions, as well as ushering in the time when the further ethical questions implicit in genetic advance will also have to be faced. Recombinant DNA technology, which I discussed in the last chapter, is one such technique. The other major ones are *in vitro* fertilization, cloning, artificial insemination by a donor (A.I.D.), and therapeutic abortion.

The technique of *in vitro fertilization* is foundational to the whole area of genetic control, because it involves the fertilization of human eggs outside the body. Once it has proved possible to interfere with the early stages of human development in this way, the remaining far more dramatic developments will be accomplished, given the necessary time and resources.

In 1966 Roberts Edwards, of Cambridge University, demonstrated how eggs extracted from human ovaries could be cultured in the laboratory ('test-tube') with the development of ripe eggs. This was followed in 1969 by the fertilization of such eggs using human sperm, and the subsequent, apparently normal, development of the fertilized eggs. Taking these techniques further, Edwards reported that his group had been successful in taking some eggs as far as the blastocyst stage, by which time the fertilized egg had divided into as many as 60-100 cells. This is true 'test-tube fertilization' and, while the blastocyst represents a very early stage in development, it is sufficiently advanced for implantation into a woman's uterus and subsequent maturation. In 1976 it was reported that an embryo fertilized in the laboratory had developed for 13 weeks in a woman's uterus while, in 1978, the first 'test-tube baby' was born amid a flurry of excitement at Oldham, in England. This was the result of the combined work of Dr. Robert Edwards and a gynaecologist, Mr. Patrick Steptoe.

In vitro fertilization has a number of potential applications. These include the alleviation of infertility brought about by a blockage in the wife's uterine tubes; the sexing of embryos, the importance of which emerges in sex-linked genetic disorders which can be avoided by using only female blastocysts; and the modification of the embryo itself to mask certain genetic diseases. At present, by far the major indication is infertility.

Edwards[1] and his colleagues recognize as their chief aims the needs of their patients and the medical well-being of any resulting children. They see no objection to discarding those foetuses with genetic abnormalities on the grounds that this is preferable to either aborting affected foetuses or producing handicapped children. On the other hand, Edwards insists on retaining the distinction between laboratory and clinical studies. In his estimation, clinical studies are justified as in this instance the embryos are the by-products of abortions and are employed in attempts to overcome infertility. By contrast, laboratory studies are based on the premise that

[1] Edwards, R. G. 'Judging the social values of scientific advances' in Birch and Abrecht, *op cit.*, pp.6-19. See also 'Aspects of human reproduction' in Fuller, W. (ed.), *The Biological Revolution* (Anchor Books, New York, 1972) pp.128-144.

human embryos can be deliberately initiated in the laboratory and subsequently destroyed, implying that these stages of life are expendable and can be subordinated to experimental ends.

Objections to *in vitro* fertilization touch on fundamental issues in the genetic advance debate. One of the most general of these highlights the nature of conception. Instead of being intimate and personal, this technique converts it into an impersonal process. It reflects in a fairly extreme form the fact that human procreation is coming increasingly under conscious purpose and control.

While arguments of this type can readily assume emotive overtones, they require serious consideration. Rejection of technological intervention in human procreation simply because it is technological finds no support in Christian terms. Man's creation in the likeness of God and his status as God's viceregent place upon him an obligation to develop technological expertise. Hence human procreation, like so many other facets of human life, has come under human aegis and consequently control. This being so, the fulcrum of the technological debate shifts to the uses being made of this control.

In vitro fertilization makes possible the alleviation of infertility by intervening in procreation outside the body, as opposed to within the body via hormonal or related measures. As such, the manner of the technological intervention has been altered, although in terms of the procreative event as a whole *in vitro* fertilization does not appear to introduce any radical procedures. It is simply an extension of therapeutic intervention as traditionally conceived.

It may be argued, though, that because the manner of intervention has changed, its essence has also been transformed. This is because it is the foetus which is now the subject of the intervention, rather than the mother. Professor Paul Ramsey[1] has written extensively on this point to the effect that experiments on foetuses are unethical because foetuses are unable to consent to the procedures. Basic to this attitude is the possibility of harm to the foetuses and, coupled with this, the objection that a hypothetical or unborn child is being submitted to a dangerous procedure. Ramsey has expressed this issue in various forms: the possibility of damage to the *in vitro* fertilized baby cannot be completely excluded, while mishaps must be discarded. Furthermore, such hazards are being imposed non-therapeutically on the child-to-be without its consent. As this foetus has been artificially conceived, the risks to which it is exposed are man-made and. therefore, need not have arisen in the first place.

Even the plight of an infertile couple does not, according to Ramsey, justify *in vitro* fertilization as 'it is not a proper goal of medicine to enable women to have children . . . *by any means*—means which *may* bring hazard from the procedure, *any* additional hazard, upon the child not yet conceived.'[2] In arguing this way, Ramsey is stressing the importance of

[1] Ramsey, P. *Fabricated Man* (Yale University Press, New Haven, 1970).

[2] Ramsey, P. 'Shall we reproduce? I: The medical ethics of *in vitro* fertilization' in *J. Amer. Med. Assoc. 220,* 1346-1350 (1972); 'Shall we reproduce? II: Rejoinders and future forecast' in *J. Amer. Med. Assoc. 220,* 1480-1485 (1972).

the individual and an offshoot of this, namely, that the child-to-be cannot volunteer. For Ramsey this in turn implies that the child-to-be cannot be exposed to any risks inherent in *in vitro* fertilization, so that the solution for the parents must be to forego bearing children.

Ramsey contends that the treatment of infertility should be by direct surgical therapy and not by *in vitro* fertilization. When employed in this manner, *in vitro* fertilization is taken out of the domain of medical ethics and into the realm of biological manufacture. This is because it is being used to meet human *desires* as opposed to treating human *maladies.* Even though we may not be prepared to go as far as Ramsey's argument takes us, there can be little doubt that this use of *in vitro* fertilization is an illustration of the employment of a *medical* technique as a *social* tool.

Ramsey's emphasis[1] on the individual as the goal of medical treatment is a welcome one, although his limitation of medical concern to the patient alone and to actual illness alone raises profound questions about the nature of personhood. Medicine frequently takes account of non-biological questions and of the future welfare of future human beings. While it is true that both these aspects of medical practice can be misused and, if allowed, may open the door to the sociological manipulation of medicine, it is hard to see how medicine can be concerned with the welfare of the whole person and yet ignore the contributions of these areas.

But what about the prospective foetus ? The welfare of this individual-to-be cannot be considered in the same way as can that of already existing individuals. Frequently the welfare of a prospective child is beyond many facets of detailed planning, quite irrespective of the method of conception. This is a part of the unknown into which we, as humans, are moving, and it simply illustrates one dimension of our finiteness.

Medical practice, particularly genetic counselling, affects yet-to-be-born individuals, affecting their probability of being born, and their genetic constitution. The difference between this situation and that engendered by *in vitro* fertilization is the difference between the indirect and direct consequences of human actions. The moral responsibility is comparable in the two categories. *In vitro* fertilization ushers in an additional course of decision-making; this, in turn, increases the tensions latent within the human condition and also the scope of treatment available to future generations.

A more fundamental consideration is that *in vitro* fertilization may debiologize procreation, marriage and the family. Dr. Leon Kass has argued persuasively in this area. He writes: '*Human* procreation is begetting. It is a more complete human activity precisely because it engages us bodily and spiritually, as well as rationally . . . What is new is nothing more radical than the divorce of the generation of new life from human sexuality and ultimately from the confines of the human body.'[2]

1 For an informed critique of Ramsey's ethical position on genetic engineering, see Curran, C. E., *Politics, Medicine and Christian Ethics,* (Fortress Press, Philadelphia, 1973) pp.164-219.

2 Kass, L. R. 'New beginnings in life' in Hamilton, M. (ed.), *The New Genetics and the Future of Man* (Eerdmans, Grand Rapids, 1972) pp.15-63.

To move away from the physical and sexual is to deprive procreation of its human connotations, because it no longer involves the diversity of factors constituting human love. This has implications for the family as a biological unit, because transfer of procreation to the laboratory undermines the justification and support which biological parenthood gives to the monogamous marriage. McCormick[1] has argued that it is in the family that we learn to become persons, experiencing the basic form of human love and caring and learning to take possession of our capacity to relate in love. To undermine the family, therefore, would be to compromise the ordinary conditions of our growth as persons.

Such a result will only ensue, should the procedure become commonplace. The central issue is that the procreative process be regarded as an indivisible entity, while stress needs to be laid on the overall situation of the husband and wife. The potential of *in vitro* fertilization lies in its ability to rectify a missing element in the union of husband and wife, so that it may become a legitimate means of healing in certain situations. Nevertheless, this does not justify its use as a way of bypassing the normal means of human procreation in the absence of a therapeutic rationale. A technical form of reproduction is inferior to one involving the bodies and personalities of two individuals, and should be resorted to only when the other fails. There is no likelihood that either *in vitro* fertilization or its technical off-shoots will become a panacea for man's ills, because these technological innovations are as fraught with dilemmas as are other aspects of the human predicament. They may become socio-therapeutic tools, however, and the implications of this need to be squarely faced.

The second technique I want to consider is *cloning,* which may be the best-known of all prospective genetic techniques. Unquestionably, it is one of the most provocative and forbidding of the techniques, and yet it is also the least realistic.

Cloning involves the production of carbon-copies of individuals. It is this technique that will allegedly lead to the production of endless streams of our great men, petty tyrants, and ordinary individuals. Fathers will be able to generate unlimited sons genetically identical to themselves and, similarly, mothers and their daughters.

Cloning is a sexual reproduction, with the result that the new individual or individuals are derived from a single parent and are genetically identical to that parent; hence the exact copies. Cloning is brought about by the removal of the nucleus from a mature but unfertilized egg and its replacement by the nucleus of a specialized body cell of an adult organism. This was first accomplished in 1961 by Dr. John Gurdon in Oxford. The egg with its transplanted nucleus proceeds to develop as if it had been fertilized, and produces an adult organism that is genetically identical to the organism serving as the source of the transferred nucleus. In this way it is possible to produce an unlimited number or clone of identical individuals. Cloning has been successfully accomplished in experiments involving

[1] McCormick, R. A. 'Genetic medicine: notes on the moral literature' in Marty, M. E. and Peerman, D. G. (eds.), *New Theology No. 10* (MacMillan, New York, 1973) pp.55-84.

frogs, salamanders and fruit-flies, although many technical feats will have to be accomplished before human cloning becomes a realistic option.[1]

Dr. Joshua Lederberg see the potential value of this technique in simple terms: 'If a superior individual—and presumably genotype (genetic constitution)—is identified, why not copy it directly, rather than suffer all the risks, including those of sex determination, involved in the disruptions of recombination ?'[2] Arguments in favour of this technique lie in the ease with which organ transplantations could be carried out between the members of a clone, although advances in immunology are a surer way of achieving this end. Cloning would also prove a way of selecting the sex of a child as all members of a clone are of the same sex, but it is difficult to take this seriously when far less spectacular means of identifying the sex of a foetus are already available. For most medical purposes these alone are quite adequate.

The essence of a critique of cloning revolves around the question of whether the postulated advantages outweigh the disadvantages. Both major arguments in its favour concern processes which, with little doubt, will be accomplished in the near future using far less dramatic techniques. Cloning, in order to achieve its goals, necessitates altering the whole of the human person and not just entities of that person such as tissues or organs. It is reductionism taken to absurd lengths.

Although cloned individuals would have much in common—in genetic terms—with identical twins, there would be a number of radical differences between them. Of these, the paramount one would undoubtedly be that the members of a clone had been produced *in order to resemble the characteristics of someone else.* Their value in the eyes of their progenitors would lie in the extent to which they replicated a previous person, and thus manifested once again certain features of that person. They would not be brought into the world in order to develop as unique individuals and to *be* themselves, but to perform specified functions and develop specified traits.

Cloning also brings us face-to-face with the nature of the family and of marriage. It demands that, under *normal* circumstances, sex is eliminated, marriage is divorced from parenthood and, one imagines, child-bearing is separated from the home. The separation of sex and reproduction as a normal procedure may well illustrate the fragility of interpersonal relationships when these cease to have any social meaning. The prospects for children brought up in such a society require careful exploration, as much—dare one say—as that given to the prospects for genetic engineering. Ramsey's cautions are well-founded when he writes: 'the thought and possibility of cloning a man may provide the truly propitious opportunity

1 Perhaps the principal reason why David M. Rorvick's account of the first human to be cloned (*In His Image* (Nelson, Melbourne, 1978)) was generally considered a hoax was due to the claim that enormous technical strides had been made in far-too-short a space of time. Cloning, as illustrated in this book, highlights human self-centredness, pride, arrogance and disdain for the well-being of others—characteristics typical in its misuse.

2 Lederberg, J. 'Experimental genetics and human evolution' in *American Naturalist 100,* 519-531 (1966).

for acquiring the knowledge that the link between sexual love and procreation is not in us a matter of animal consequence only, but of truly human and personal import . . .'.[1]

Up to this stage in the discussion of genetic techniques, my concern has been with techniques that are either being developed at present or remain future possibilities. With the third technique to be discussed, *artificial insemination by a donor* (A.I.D.), we move into the realm of current practice. What is perhaps particularly surprising about A.I.D. is the extent to which it is current practice. Relatively little is heard about it and yet probably well over 10,000 artificial inseminations a year are carried out in the United States, and at least 2,500 a year in a country with a small population like Australia. These figures suggest that A.I.D. is increasingly becoming an accepted means of circumventing the prospect of a childless future for many infertile couples, especially as adoption becomes increasingly difficult.

Sperm can be stored in liquid nitrogen at very low temperatures, thereby maintaining indefinitely their viability. Consequently, sperm from numerous donors can, in this way, be maintained with the building-up of sperm banks. These banks mean that sperm can be not only kept, but also coded according to donor characteristics. As a result, an extensive choice of donor sperm is made available enabling the closest possible match in physical and mental characteristics between the husband and donor. Each donor is carefully screened to ensure that there are no mental or physical conditions to render him unsuitable as a donor. Not surprisingly, therefore, A.I.D. births have a lower-than-expected number of abnormalities.

An immediate problem is to decide who is entitled to receive A.I.D. The whole-hearted consent of both husband and wife is generally required, and as far as possible should be insisted upon. An equally pressing problem is the selection of donors. It is relatively easy to choose medical students, as has frequently been done, and while careful screening is carried out this cannot be exhaustive. Assuming that eugenic ideals are discarded, the most that can be expected of donors is that they are responsible citizens with acceptable I.Qs.

The legal issue in this debate revolves around the question of whether A.I.D. constitutes adultery. Although there has been some doubt in English law on this latter point, it appears that A.I.D., even apart from the husband's consent, does not amount to adultery by the wife. However, as English law now stands, the child undoubtedly is illegitimate.

Arguments adduced in favour of A.I.D. regard the source of the reproductive cell as far less important than the love, care and nurture of children. In terms of this argument, children constitute the essential element of parenthood, and if children do indeed constitute the sole criterion by which reproduction is judged the means by which they were conceived becomes irrelevant. Such undue emphasis on the begetting of children may itself highlight the fragility of the family unit and may, therefore, over-ride the essence of family relationships. The plight of the childless couple, while

[1] Ramsey, P., *Fabricated Man, op. cit.*

calling for sympathy and understanding, needs to be viewed alongside the stability of the couple as marriage partners. If A.I.D. is a factor liable to disrupt the marriage relationship, the conclusion must be drawn that a couple does not possess an absolute right to have children.

A.I.D. introduces into the family unit only half an outsider, namely, a child carrying the wife's genes but not the husband's. In this regard the child is more a part of the family than is the adopted child. However, in order to accomplish this, a biological bond between the husband and wife has been severed. A.I.D. involves the radical separation of marriage and parenthood, a separation also illustrated by cloning.

Once man and, in particular, procreation are subdivided into a biological aspect that can be technologically manipulated, and a personal aspect that is above such manipulation, the integrity of human parenthood is lost. In specific terms, therefore, a wife who gives birth to a child by A.I.D. without her husband's consent appears to have violated the essence of the marriage.[1] This may not amount to adultery, but it is an offence against the exclusive nature of her marriage vows.

An issue repeatedly raised by the preceding discussion has been that of *therapeutic abortion.* Ethical difficulties abound because there is no certainty that the foetus considered for abortion will definitely be affected. There is a statistical risk of defect, which may be as high as 50 per cent in, say, haemophilia although lower for many other disorders. The resulting decision involves destroying healthy foetuses in order that a smaller number of defective foetuses will not be allowed into the world.

Except for those who, at the one extreme, are totally opposed to abortion under any circumstances, and for those who, at the other extreme, contend that every child has the right to be born with an adequate hereditary endowment, a decision involving foetus, parents and siblings has to be made. The decision to deprive a foetus of its potential for life is a weighty one, and may depend on a decision that the disadvantages of its birth are greater than the benefits. A decision of this order needs to take account of a range of factors including the severity of the genetic disorder and the quality of life the afflicted child will have; the physical, emotional and economic impact on the family and society; the reliability of the diagnosis; the increase in the load of detrimental genes in the population should the afflicted individual subsequently produce children. There are no easy answers. For instance, on the positive side a defective child may be able to compensate for its defect; on the negative side the unity and health of a family unit may be jeopardized by extreme deformity.

I would not advocate a hard-and-fast position on therapeutic abortion, although I would err on the side of reverence for human life. A decision regarding therapeutic abortion should express the wholeness, freedom and responsibility of individual and family decision-making. Above all, society even at a cost to itself should hesitate to restrict the freedom of an individual wishing to bear a defective child.

[1] See Anderson, J. N. D., *Issues of Life and Death* (Hodder and Stoughton, London, 1976) p.48.

4. GENETICS AND THE HUMAN PREDICAMENT

In this chapter I shall turn to a few central issues for Christians confronted by one or more of the dilemmas raised by contemporary genetics.

1. The nature of man

Man is a being created *in the image* and *after the likeness of God,* from which it follows that he is a rational and morally responsible being with freedom of moral choice. Moreover, he is a creature with whom God can communicate and who can himself enjoy a personal relationship with his creator.[1] Any decisions concerning human beings, therefore—their genetic structure, their future existence or otherwise, their freedom to marry and bear children—should start from the fundamental premise of their God-relatedness. To ignore this is to accept some form of reductionistic starting-point.

A further consequence of man's God-relatedness is the dominion over the created order given to man by God. Man is responsible for the created order and is to exercise his power in accordance with God's moral nature. He is obligated, therefore, to use his mind in the exercise of this dominion, the corollary of this being that he is unjustified in abusing his biological, and especially his genetic, inheritance, either through indifference or blatant disregard of the consequences. General as this principle is, it is the hallmark of Christian responsibility in genetics, emphasizing as it does the regard to be displayed by one human being for the person and freedom of another. To rob a man of his freedom is to rob him of responsibility and hence of a central facet of what it means to be human.[2] This applies at the level of groups as well as at the individual level.

Related to this emphasis is the tenet that man is both *in* nature and *over* it; he is both a spiritual being and a biological one.[3] In short, he is a unity. From this it follows that man cannot be dissected, either by genetics or theologians, into their favourite separate elements and still maintain his identity. This holistic approach has much to say with regard to genetic procedures, the incipient tendency of which is to subdivide man into his smallest genetic components. Genetic techniques. however, have the potential for achieving wholeness and in this sphere they are to be encouraged, being manifestly in line with the biblical perspective of man's unity. This is the basis of the therapeutic utilization of genetic procedures. the aim of which is to deepen a holistic vision of the human situation.

2. The improvability of man

As this discussion has made abundantly clear, it is impossible to consider the nature of man without asking whether man can be improved upon, biologically, psychologically and spiritually. If so, the ever-increasing technological armamentarium of the geneticists should prove a potent factor in bringing about improvement.

1 Anderson, *ibid.* pp.20, 21.

2 Hughes, P. E., *The Control of Human Life* (Presbyterian and Reformed Publishing Company, Nutley, New Jersey, 1973) pp.3ff.

3 Kidner, F. D., *Genesis* (Tyndale Press, London, 1967) p.50.

Professor Donald MacKay[1], in tackling this issue, suggests that improvement would mean an enhancement of people's capacities to relate better to God and to one another. In drawing out this principle, he distinguishes between contentment with the unalterable and complacency with the alterable, and lays down as a fundamental criterion the question: What will make the most acceptable possible offering to God? For instance, to produce improved human specimens at the expense of family relationships, spiritual dimensions, or human attitudes such as compassion, self-esteem and individual dignity, is to place greater value on biological improvement than on those relationships considered by Christians as fundamental to the human condition.

Bernard Häring[2], by contrast, argues that since man's biological nature is entrusted to his freedom and wisdom, he is the steward of his genetic heritage. This enables him to plan his genetic future (as and when the appropriate techniques become available) and so, Häring contends, man may be entitled to accelerate hominization by direct improvement of his genetic heritage. It appears Häring is able to hold this position on the assumption that creation, including man himself, is an unfinished work that calls for man's co-operation in bringing it to 'greater perfection'. Hence man is, according to Häring, called to become an even better image of God.

The tenuousness of Häring's position is evident even from his own writings. He admits that acceptable criteria must be found to minimize the possibility of unrealistic and irresponsible interventions, while he quotes with approval Karl Rahner's question whether the intelligence that can be genetically influenced is the kind of intelligence most needed. Implicit within this is the fundamental distinction between improvement of biological quality and improvement of moral qualities such as wisdom, moral responsibility and altruism. Häring's advocacy of biological quality control in no way contributes to this debate, and he is forced to make a further distinction between improvement of man's genetic heritage and change of the human species. By the latter, I take him to mean extreme forms of genetic manipulation that radically interfere with genuine human intercommunication. In practice, therefore, Häring would probably allow only limited self-modification; but the difficulties of his position appear to outweigh the postulated benefits of this small degree of self-modification.

Resolution of this issue necessitates a return to the creation precepts and also consideration of the incarnation. Not only is man made in God's image, but the Son of God took to himself human identity, thereby revealing the image of God in man. An outcome of this is that Christ, who was one with God from before the creation, became a real man at his incarnation without in any way ceasing to be God.[3] Not only this, but when Christ ascended he maintained his manhood—albeit glorified and exalted. In

[1] MacKay, D. M. 'Biblical perspectives on human engineering' in Ellison, C. W. (ed.), *Modifying Man: Implications and Ethics* (University Press of America, Washington, 1977) pp.67-95.

[2] Häring, B., *Manipulation* (St. Paul Publications, Slough, 1975) pp.64, 183ff.

[3] Mascall, E. L., *The Importance of Being Human* (Oxford University Press, London 1959) p.22.

remaining man, Christ has bestowed upon the human race, and human beings in their *present* form, an unequalled value. Consequently, individuals as we now know them are of such value that we are to be satisfied with the type of body, brain and genetic inheritance they currently possess. There is room neither for the fashioning of a radically new type of human being, nor for the manipulation of individuals in the hope of producing improved and perhaps novel beings. On the other hand, recognized defects in individuals demand treatment in accordance with therapeutic expectations. This is reminiscent of a distinction made by MacKay, to the effect that compassion for our fellow men is not the same as ambition for them.

3. The status of individuals

In the sight of God individuals are important, from which it follows that they are to be respected simply because they are human beings created in the image and after the likeness of God himself. Each individual person is an end in himself, so that the interests of a person—whether a patient or even an existing foetus—should not readily be subordinated to other interests, such as the good of humanity, the future of the human race or the progress of medical science. Although it may not always be possible to draw clear-cut distinctions, particularly when dealing with foetuses, the claim of the individual is paramount. This may take the form of protecting a carrier from discrimination, or allowing a patient the freedom to remain ignorant of a suspected diagnosis, or enabling a subnormal child to live. These may not be absolute claims, but when the status of an individual is at stake they demand serious consideration.

Human life is precious to God, as revealed by his creation and redemption of mankind. To undervalue human life is to underrate these facets of God's character. In particular it negates the incarnation with its suggestion that human life is unworthy of Christ's high estimation of it. This emphasis on the individual raises a further principle, that we are not to create life irresponsibly either by natural process within marriage or outside of it, nor by laboratory experiment.[1] Both natural and artificial reproduction are to be judged by circumstances affecting the prospective child, parents, siblings, and possibly also society. Man's dependence upon God, the value of individuals and the respect shown by Christ for mankind are essential considerations if a biblically-oriented viewpoint is to emerge.

Closely related to this discussion is an allied one, namely, the *dignity of man.* This refers to the dignity bestowed upon man by God in creation and redemption. Professor Bernard Ramm recognizes in this dignity the only point of leverage in ethics, because according to it man is to be treated as man in his God-given dignity whether he is sullied or redeemed. The immense significance of this principle derives from the fact that: 'the dignity of man rests not upon his functional ability, but rather upon the fact that God loves him, that he was clearly purchased, that Christ died for him, and that, therefore, he stands under the protection of God's eternal goodness.'[2] The implication of this is that those who have no functional value in society,

1 Dunstan, G. R., *The Artifice of Ethics* (S.C.M. Press, London, 1974) p.70.
2 Ramm, B. 'The common call for human dignity' in Ellison, *op. cit.,* pp.135-139.

the genetically disadvantaged and the mentally retarded, still have an 'alien dignity'—a history with God and the sacrifice of Christ on their behalf. For Helmut Thielicke, only in his alien dignity is there any security in a world where manipulation, genetic and otherwise, is a factor within society.[1]

4. Man's finiteness and sinfulness

Man is limited, being a creature in a God-ordered and God-sustained world. His creatureliness implies that he is limited by his own biological make-up. He must operate within a particular natural and social framework, inherent within the Creator-creature relationship. Man is not a god and there is no radical way in which he can transcend his creatureliness. To speak, therefore, of man as if he had himself become a creator is grossly misleading, specially when it is implied that man now possesses the ability to change the biological rules controlling his genetic inheritance. Hopes of this order fail to analyze seriously the extent of man's genetic knowledge, and flounder on determining the goals for man's self-modification.

More often than not, the reasons for biological modification are grounded far more in what man can *do* than in what is *needed.* Debates about man's genetic future invariably centre on practical procedures, the desire being to transcend man's creatureliness in order that he might cope biologically with what he has created technically. The means themselves have become the model of what man's ends should be, so that man may lose the distinction ot being an end in himself.[2] This may reflect an unwillingness to face up to the challenges implicit in human existence and to exploit the potentialities of what it means to *be* human. By resorting to technological answers to human questions, it is all too easy to turn away from the realities of human existence, with its freedom, fear and pain, and turn instead to an idealized world of biological perfection.

The sinfulness of mankind is a related, but different, aspect of the human condition. This illustrates the polarization between what man *is* and what he can *do,* emphasizing the dichotomy between man's efforts to attain wholeness and the incipient tendency towards destructiveness. Examples of this are found in the bringing-to-term of a malformed foetus, or in the enabling of individuals with certain genetic abnormalities to survive to reproductive age thereby passing on the deleterious genes to their offspring. In the first instance, an unfit individual who would not otherwise have survived is enabled to do so with widespread ramiffcations for himself, his family and society. In the second example, the consequence is deterioration of the genetic pool.

Biological solutions are incomparable in attacking essentially biological problems; they are, however, ill-fitted for solving problems rooted in the nature of man. Progress in genetics cannot dispense with the basic conflicts of human existence, simply because man is not solely a biological being. He is a unity—biological, aesthetic, religious and personal. And

[1] Thielicke, H., *The Doctor as Judge of who shall live and who shall die* (Fortress Press, Philadelphia, 1976) p.41. See also his *Between Heaven and Earth* (Clarke, London, 1967) pp.160ff.

[2] Thielicke, *The Doctor as judge of who shall live and who shall die, op. cit.,* p.37.

if, as the biblical writers assert, his relationship to God has been marred, genetic manipulation can have no effect on this relationship or its consequences. Fallen man is man in conflict—with himself, his neighbours, his environment, and his God. Far more basic than his need of genetic manipulation is his need of God's forgiveness, although these are in no sense mutually exclusive needs. Both will benefit man, although at different levels of his existence and to different degrees. While redemption is essential to man's standing in God's sight, genetic manipulation effects only the quality of man's biological life. While the latter cannot be ignored, neither must it be equated with the former.

Yet another aspect to this discussion is that man in his sinful state will misuse the possibilities opened up by genetic manipulation. This is not an all-or-nothing phenomenon, as the good and the evil uses to which it will be put are intermingled. Consequently, the ways in which genetic manipulation may mould man's life will reflect man the sinner living in a world tainted by sin. In no sense, however, does this negate its beneficial aspects within a biological framework.

5. Dehumanization

To dehumanize a fellow human being is to treat him as an object rather than as a being like oneself. It is to deny him his full human connotations of freedom, reason, body and emotions, suppressing one or more of these from his humanity.[1] This may happen when the origin of the child is removed, for no good therapeutic reasons, from the sphere of marital love. Alternatively, it may occur when the principal criterion for human survival, becomes biological quality, thereby ignoring the non-biological traits of human existence and hence the wholeness of the human condition. This in turn, introduces the concept of manipulation because any treatment jeopardizing the vision of human wholeness is manipulatory.

Basic to an understanding of manipulation are the related criteria of *wholeness* and *freedom.* Inherent within human wholeness is the desire to see man as a unity, and to encourage the development of the various facets of this unity. The over-emphasis of a single dimension of human life, therefore, is to move away from wholeness, as is the neglect of any particular dimension. Genetic treatment has an important role to play in the attainment of wholeness, although dangers are also evident if a single dimension of human existence is isolated and magnified to the detriment of the person in all his relationships. It is because of considerations like these that a child with Down's syndrome, for instance, may bring out essentially human characteristics in those closest to him in spite of his relative lack of biological quality. Conversely, the attainment of biological quality by aborting an afflicted foetus may, in some situations, be at the expense of compassion, love, and sacrificial self-giving. In this example, however, the pros and cons of the individual circumstance need careful examination before a responsible decision can be made.

Freedom to develop our potential as human beings is basic to the human situation. Häring writes that man 'has to gain the inner freedom to be, to love, to adore.'[2] What this means in the genetic realm is that each person

[1] Häring, *op. cit.*, p.50.
[2] *Idem.*

born into the world must have the opportunity, as far as possible, to develop in this direction. Any procedure that may limit this, be it cloning or stringent limitations on a person's freedom of choice on genetic grounds, becomes manipulatory.

Dehumanization is an ever-present possibility: choosing the superficial characteristics of our children, routinely conceiving apart from sexual intercourse, isolating child-bearing from the family unit, and transferring pregnancy to the laboratory as a routine procedure. Each of these, when performed out of choice and for no overbearing medical reasons, reduces a human activity to an inhuman, impersonal technique. The intimate cohesion of body, mind and personality is lost, and human existence is demoted to a robot-like mechanism. When employed in this context, these techniques contravene the basic biblical constraints of the significance of individuals, the unity of marriage and parenthood, and the cohesion of parenthood and childbearing.

6. Possibilities of genetic therapy

The positive side of the debate on genetic advance is seen in the possibility of using genetic techniques to enhance the meaning of human existence by reducing the burden of illness and by increasing the opportunities for deepening the richness of human experience. Most of the techniques previously discussed, with the exception of cloning and A.I.D., contain such hopes within them as long as they supplement a view of man which gives due weight to the meaning of his life, to values such as justice, freedom, compassion and love, to moral actions, and to a world view in which science helps provide a knowledge of what things mean and not simply with what they do.

The role of genetics depends on the ways in which its techniques are directed and applied by human beings. This is why the nature of man is to paramount importance in the genetic debate. Once man is assumed to be autonomous, genetic advance can readily acquire manipulatory possibilities. Conversely, as long as man is recognized as the handiwork and servant of God, who creates and redeems, genetic advance becomes a means by which God can alleviate some of the suffering and anguish of mankind.